Table of Contents

Natural and Artificial Flavors

What's the Difference?

WRITTEN BY

Josh Bloom, Ph.D.

A publication of the

AMERICAN COUNCIL
ON SCIENCE AND HEALTH

American Council on Science and Health
110 East 42nd St, Suite 1300
New York, NY 10017-8532
Tel. (212) 362-7044 • Fax (212) 362-4919
URL: http://www.acsh.org • Email: acsh@acsh.org

Publisher Name: American Council on Science and Health
Title: Natural and Artificial Flavors: What's the Difference?
Author: Josh Bloom, Ph.D.
Subject (general): Science and Health
Publication Year: 2017
Binding Type (i.e. perfect (soft) or hardcover): Perfect
ISBN: 978-0-9910055-9-8

Acknowledgements

The American Council on Science and Health appreciates the contributions of the reviewers named below:

Rhona Applebaum, Ph.D.
Executive (retired)
Food Association
New York, NY

Martin Di Grandi, Ph.D.
Associate Professor of Chemistry
Department of Natural Sciences
Fordham College at Lincoln Center

Joe Schwarcz, Ph.D.
Professor of Chemistry
McGill University, Montreal

Michael D. Shaw, Ph.D.
Executive Vice President
Interscan Corporation
Reston, VA

Christoph W. Zapf
Associate Director, Medicinal Chemistry
Nurix, Inc.
San Francisco, CA

1
Introduction

Of the many misconceptions used in the "natural vs. artificial" narrative, two stand out: (1) That artificial flavors are inherently less healthy than their natural counterparts, and (2) that a flavor chemical obtained from a natural source is either different or superior to the same flavor chemical produced in a laboratory or factory.

Together, these beliefs represent a cornerstone of the natural movement. As pervasive as this mindset is among consumers of "organic" and "natural" goods, it violates simple laws of chemistry.

Not only is this belief false, there are actually times when the opposite can be true. For example, an artificial flavor made in a lab will typically be approximately 100 percent pure, while that same flavor that is obtained from a plant will not. A natural version will contain other chemicals, which make up the flavor of the food, and some of these natural chemicals can be toxic, or even carcinogenic, while an artificial flavor won't contain these substances. Some of the chemicals that comprise the mixtures of natural flavors or scents have even been characterized by environmental groups as dangerous. But as you will see, they are nothing of the sort.

The truth is multiple chemicals that make up natural flavors in a piece of fruit are not harmful. They are not toxic in natural foods for the same reason they are not toxic in artificial ones — they can't be. As wisely codified by Paracelsus, the noted 16th century scientist often considered to be the founder of modern toxicology, the dose makes the poison. Or none of us would have survived this long.

Yet, food marketers unabashedly exploit natural-versus-artificial fallacies. The trend began in health food stores but it has spread throughout the entire industry. "No artificial flavors" is prominently displayed on the labels of one product after another, including macaroni and cheese, cookies, candy bars and jelly beans.

There is, of course, is no obvious health downside to consumers who choose products that are advertised as containing "no artificial flavors." They will probably pay more to get something that is just made by a different process. It may or may not taste the same, but that's it. There is harm here in continued dissemination of factually incorrect science to Americans, which indirectly assaults all of us, but most important is the manipulation of those who can't afford to choose to overpay for foods and goods that offer nothing more than imaginary benefits. It is these people pressured by marketing claims — that any product without a natural sticker is more dangerous — who may come to think that they're bad parents if they choose conventional products for their kids.

Environmental groups have spent hundreds of millions of dollars trying to convince people that there are harms associated with exposure to trace levels of chemicals, especially those added to food. This marketing chicanery of the food industry is so pervasive that it perpetuates an irrational fear of chemicals, and this fear has a cascade effect on public acceptance of science as it pertains to quality of life.

Consumers should always have the right to choose whatever products they prefer, but when this "choice" is built upon scaremongering a scientific fallacy, it's not a choice at all. It is an *apparent* choice, not a real one, all thanks to faulty science.

2
A chemical is a chemical, no matter its origin

The essence of the disconnect between legitimate science and erroneous claims about chemistry and chemicals is the widespread, but incorrect, notion that natural and synthetic are two distinct classes of chemicals. That is, a chemical's safety, nutritional value and flavor depend upon its origin.

This lies at the heart of some environmental group's fundraising tactics. The use of "celebrity science" — the dissemination of misinformation by those who command attention solely because of their celebrity status — is a powerful tool. Whether scientifically misguided or intentional, celebrities can use their status to reach a disproportionate share of the public, enabling them to send confusing or outright false information to millions who may lack the scientific acumen to question what they are being told.

Although hardly alone, the amateur food "expert" Vani Hari, who calls herself "The Food Babe," may be the worst offender. Hari champions beliefs such as "I won't eat anything that I can't spell," as if her spelling abilities have any bearing on the merit (safety and quality) of a chemical, food or food additive. Hari may be doing wonders for her bank statement, but she is doing an enormous disservice to the public by spreading her foolish claims along with the profoundly antiscientific message that accompanies them. Likewise, in her place you could insert Gary Null, David "Avocado" Wolfe, Mike Adams (aka "The Health Ranger") or Joe Mercola, D.O. and the message would be more or less the same.

According to Hari's eat-spell "test," cyanide should be perfectly acceptable to consume, while (5R)-5-[(1S)-1,2-dihydroxyethyl]-3,4-dihydroxyfuran-2(5H)-one (vitamin C) should not. Likewise, chlorine — one of the first chemical weapons ever used during wartime —passes muster, while (2E)-3-phenylprop-2-enal (cinnamon) does not. The absurdity of this logic is evident yet the "chemicals are bad" mantra endures with a little extra, but unneeded, help from Hari.

And this mindset is nothing but a mantra, not anything real. Chemicals are chemicals, and they all have different properties, none of which depend on spelling, something that anyone with even the most rudimentary knowledge of science will know.

The damage that the "natural pushers" do may *seem* trivial, but it is not. Their misinformed or intentionally-deceptive message confuses people by drawing an imaginary boundary between natural and artificial, whether it pertains to foods, colors, scents, or flavors and even drugs.

Science loses to marketing

For example, Joe Mercola, D.O., a supplement uber-salesman, helps spread the same phony scare[1] when discussing the chemical, diacetyl, stating, "Research shows diacetyl has several concerning properties for brain health and may trigger Alzheimer's disease." What Mercola conveniently omits is that diacetyl naturally exists in any number of foods[2], including butter, beer, wine, cheese, coffee and yogurt.

He also manages to get two things wrong[3] about the same chemical: "Many companies who manufacture microwave popcorn have already stopped using the synthetic diacetyl because it's been linked to lung damage in people who work in their factories." By highlighting "synthetic" Mercola acknowledges that the chemical diacetyl is a naturally occurring flavor, but then implies that "synthetic diacetyl" is somehow more harmful than what occurs in foods. That's profound ignorance of both chemistry and biology. Or, perhaps Dr. Mercola actually knows some science, but also knows that distorting the truth and spreading fear is a better business model.

3
The fundamental differences between natural and artificial flavors

Natural flavors are typically complex mixtures of chemicals derived from plants or fruits. In many cases there will be one predominant flavor chemical, as well as dozens, or even hundreds of other components. It is this complex mixture that gives natural extracts a richer, more complex flavor. But it is usually the predominant flavor chemical that will be identified by someone's sense of taste or smell.

By contrast, an artificial flavor is synthesized from other chemicals rather than being extracted from a natural source. Artificial flavors usually contain only a small number —often just one — of the same flavor chemicals found in the natural extract, but lack the others so they cannot precisely duplicate the flavor of the complex mixture. So, while someone tasting an artificially flavored food will be able to identify the principal flavor, it may seem bland or taste like it is "missing something." Some are better than others, so we'll discuss a few. Vanilla will be our first test case because it's relatively simple and many people like it.

4
Vanilla

Flavor

As is shown in Table 1, both natural and artificially flavored vanillas contain the same principal flavor chemical, vanillin. But the bean extract contains three other major components, vanillic acid, 4-hydroxybenzoic acid, and 4-hydroxybenzaldehyde, which account for 17 percent (by weight) of the flavor chemicals that make up vanilla.

Although none of these chemicals smell or taste like vanilla, they contribute to the flavor and scent of extract of vanilla because they have flavors and scents of their own. Since both scent and taste are subjective, it is impossible to quantify how much each of these other components contribute to what people experience when they taste vanilla, which comes from beans. But some will notice a difference, and will probably prefer the natural flavoring for this reason.

Table 1
The principal flavor components of vanilla beans from Madagascar

	Flavor Chemical(s)	Amount	Comments
Vanilla Extract (1)	Vanillin (2)	82%	
	4-Hydroxybenzaldehyde	7%	Bitter almond flavor
	4-Hydroxybenzoic acid	3%	Faint nutty flavor
	Vanillic acid	7%	Creamy flavor
Synthetic Vanilla	Vanillin	100%	

Notes
1. At least 170 chemicals have been isolated from vanilla beans
2. Principal flavor of vanilla

Safety

The LD50 — the acute dose that causes 50 percent of test animals to die — of vanillin in mice is about 3,925 milligrams per kilogram of body weight of the mouse. This means that it requires about 80 milligrams of vanillin to kill a 20-gram (0.02 kilogram) mouse. If mice were little people (they aren't; this is a crude approximation) it would take 275 grams (or more than half a pound) of vanillin to be sufficiently toxic to kill half of the people who ingested it, based on an average human weight of 70 kilograms. Bakers, for instance, know that a cake recipe calls for one-half of a teaspoon of vanilla, and that amount of vanilla extract contains 0.50 grams[4] of vanillin. Therefore, you would need to eat 550 cakes — at once — to ingest the lethal dose of 275 grams of vanillin. The cakes would get you long before the vanillin did.

That's natural vanillin. So what about synthetic? Any toxicity or health threat associated with the use of synthetic vanilla will necessarily be the same as that associated with naturally-derived vanillin, since vanillin is vanillin, no matter its source.

But there is one caveat — while synthetic vanillin is just vanilla, using the naturalistic fallacy we find that natural vanillin could theoretically be *more* harmful than its synthetic counterpart, because there are many additional chemicals present. What about the three additional predominant flavor chemicals that come from vanilla beans?

Don't be concerned, natural vanillin is actually every bit as safe as its synthetic counterpart (even though it contains chemicals that "The Food Babe" can't pronounce):

- **4-Hydroxybenzaldehyde:** No significant toxicity[5]

- **4-Hydroxybenzoic acid:** No significant toxicity[6]

- **Vanillic acid:** No significant toxicity[7]

Neither vanilla extract nor synthetic vanillin presents any health risks. The only difference between the two is perceived flavor and real cost.

5
Grapes

Flavor

Unlike vanilla, there is no single principal flavor in grapes; the flavor arises from many chemicals and sugars. The differences in composition of the natural and artificial flavors of grape and vanilla are profound. While the flavor of vanilla extract is primarily due to one chemical, the flavor of grapes is the product of hundreds of naturally occurring chemical compounds.

Compared to the relative simplicity of vanilla, the mixture of chemicals in freshly squeezed grape juice is bewildering. Seven different classes[8] of chemical compounds have been identified, and each class has multiple members. Making this matter far more complex is that there are more than 10,000 different varieties[9] of grapes used for winemaking alone.

For example, Williams, et. al, identified 26 different chemicals belonging to a single class of compounds called monoterpene alcohols from Muscat grapes[10].

The complexity of grape flavor can be illustrated even by noting a small subset of these chemicals. Note their ubiquity in nature as other flavors and scents (See Table 2).

Table 2

Select terpene alcohols found in grapes. These examples make up only a partial list of the complex mixture of these chemicals in the fruit.

Chemical	Natural Occurence
Geraniol	Roses, citronella, lemon, geraniums
Myrcenol	Lavender, grapefruit, licorice, lime
Citronellol	Apricot, basil, coriander, eucalyptus
Nerol	Blood orange, currants, carrots, rosemary
Linalool	Beer, butter, celery, nutmeg

This chemical complexity makes it impossible to even characterize, let alone duplicate, the natural flavor of grapes. But there's one chemical called methyl anthranilate, which, although it is found in small quantities, is nonetheless associated with grape flavor.

Not all varieties of grape contain methyl anthranilate, but most do. The quantity of this chemical is dependent on the type of grape, as well as the environment in which the plant was grown, the time of its harvest, and the method of extraction of the chemicals from the grape.

Interestingly, the use of methyl anthranilate for artificial grape flavoring did not follow the standard pathway —isolation and identification of chemical(s) that are responsible for natural flavor in the food, followed by use of the synthetic version of the same chemical(s) in the artificial flavor. Instead, methyl anthranilate just happens to smell and taste somewhat like grape and was thus used as an artificial flavor in candy before anyone knew that it actually existed in grapes. Only later was methyl anthranilate identified as a natural grape component.

However, though grapes contain methyl anthranilate, it's only a minor component of the enormous mixture of chemicals that comprise natural grape flavor.

That is why compared to vanillin and vanilla, methyl anthranilate alone is a poor artificial grape flavor. Liu and Gallender illustrated why in the *Journal of Food Science* (1985, 50, pp. 280-282). Concord grapes were collected from five different locations in Ohio, and the methyl anthranilate content was measured. The concentration of methyl anthranilate in the grapes in this study ranged from 0.14 milligrams per liter (mg/L) to 3.5 mg/L. Even at the highest concentration, 3.5 mg/L, the reason methyl anthranilate is a poor artificial flavor becomes obvious. On a weight-to-weight basis, methyl anthranilate makes up only 0.35% of the weight of one liter of solution.

Although the numbers are not directly comparable, it is obvious that vanillin, which comprises 82% of the flavor of vanilla bean extract, is an excellent artificial flavor — one that closely approximates the flavor of the natural flavor — while methyl anthranilate is not. Grape-flavored juices, candy, and soda often taste like a "phony" grape flavor, while cookies that are flavored with synthetic vanillin taste like vanilla. The means by which the flavor is obtained (synthesis vs. extraction) is irrelevant in both cases.

As with vanilla, the chemicals in artificial grape flavor and natural grape flavor make no difference in health, which contradicts what food scare-mongering groups contend.

Safety

Naturally-occurring methyl anthranilate comprises such a small percentage of the flavor chemicals in grapes that even with the enormous quantity of grapes and grape products consumed around the world, the chances that the chemical represents a health threat is zero — whether it's used as an artificial grape flavor or is naturally present.

As stated before, the methyl anthranilate that is produced by grapes is in every way identical to that made in a factory. So like vanillin, the chemical can only be harmful if it is used in quantities that are sufficient to bring about toxicity — but that amount is well beyond the possible limits of lifetime human consumption. As with vanillin, the toxicological properties of methyl anthranilate have been thoroughly examined[11].

Animal toxicity of pure methyl anthranilate at high doses:

▸ Exceedingly low toxicity when fed to rats, mice, guinea pigs

▸ Minor skin irritation when applied to rabbit skin

▸ Not mutagenic

▸ Human toxicity of pure methyl anthranilate at extremely high doses:

▸ Eye irritant

▸ Lung and skin irritation upon prolonged exposure

▸ Can provoke an asthmatic response (rare)

Based on the toxicity profile shown above, methyl anthranilate has a clean bill of health. The chemical is far less toxic than virtually all natural drugs or chemicals (or their synthetic counterparts) we're exposed to on a daily basis.

Conclusion

Although methyl anthranilate comprises only a very small percentage of the natural grape flavor, it is nonetheless used routinely as artificial grape flavor. Its resemblance to grape flavor, while noticeable, is considered to be poor. With regard to toxicity, methyl anthranilate has an excellent safety profile, so it would be difficult to imagine any circumstance in which this natural or artificial grape flavor could in any way constitute a health risk.

6
Bananas

Here we have a natural flavor that can be *more* toxic than its artificial counterpart. But we're not scaremongers or selling an alternative product, so we can assure you it is virtually impossible.

For a food chemical to be dangerous, three conditions must be met:

(1) As is the case with any chemical, whether natural or synthetic, the flavor chemical must have inherent toxicity;

(2) The exposure (or dose) must be sufficient to cause adverse effects; and

(3) The metabolism of the chemical in the body must be slow enough to allow a buildup to toxic levels, or to produce a metabolite that is more dangerous than the chemical itself.

A natural banana meets these parameters, and if we were at Center for Science in the Public Interest our lawyers might sue banana companies to make some money over it. But bananas meet these parameters in a non-meaningful way.

On perceived taste, bananas provide a good example of an artificial flavor that lies between vanillin (an excellent mimic of the flavor of vanilla), and methyl anthranilate (a lesser quality mimic of the flavor of grapes).

Most importantly, since a good number of flavor and/or scent chemicals found in bananas have been isolated and their chemical structures elucidated, bananas provide a textbook example of the vital relationship between dose and toxicity. While bananas contain a variety of chemicals considered moderately toxic, we do not die from their consumption.

It is this paradox — some chemicals in bananas can be potentially hazardous, but they do not harm us — that makes the fruit an excellent teaching tool for debunking commonly-held myths about what the terms natural and artificial really mean, with regard to both taste and health. This is demonstrated in Table 3.

Table 3
Flavor and color chemicals in bananas

Banana Chemicals	Additional Information
E1510	Ethyl alcohol
E306	Tocopherol (vitamin E) rich extract from vegetable oils
E515	Potassium sulfate, electrolyte imbalance from large amounts
Ethyl 2-hydroxy-3-methyl-butanoate	Caramel-like odor. Minimal toxicity
Ethyl butyrate	Pineapple odor. Used as a flavor additive for orange juice.
Ethyl hexanoate	Fruity odor, component of pineapples and apples.
Ethylene	Petrochemical. Can be explosive in high concentration. Natural ripening hormone of many fruits.
Isoamyl acetate	The principal flavor of bananas. Produced by the plant or synthetically. Harmful only at very high doses
Isoamyl alcohol	"Disagreeable" odor. Minimal toxicity.
Isobutyl acetate	Flavor from raspberries, pears. Harmful only at very high doses
Isobutyl alcohol	Sweet, musty odor. Minimal toxicity.
Isobutyraldehyde	Sharp, pungent odor. Moderate toxicity.
n-Pentyl acetate	Banana-like odor. Very similar to isoamyl acetate
Yellow-brown E160a	Also known as beta-carotene, a source of vitamin A
Yellow-orange E101	Also known as riboflavin (vitamin B2)

Perhaps no chemical in bananas illustrates the confusing and incorrect uses of the terms "natural" and "artificial" better than Yellow-brown E160a, also known as beta-carotene (β-carotene), a biosynthetic precursor of vitamin A. Yellow-brown E160a is a carotenoid, a fat-soluble oil that is ubiquitous in nature. It is biosynthesized by bananas, as well as many other yellow and orange colored fruits and vegetables[12], such as carrots, pumpkins, sweet potatoes and tomatoes.

But the primary industrial use of β-carotene is an artificial color that is used to make foods, such as butter and margarine, yellow. Does that make it a natural or artificial color? Since the chemical is added to foods one could argue that it's either artificial because (a) the yellow color does not naturally appear in the food, or (b) natural, because it is found throughout the plant kingdom. It gets even more confusing if you try to create a world where natural is inherently "good" and artificial is "bad." Although β-carotene occurs in, and can be extracted from, many natural sources, the raw material in a $300 million annual market usually comes from a factory. β-carotene is typically made synthetically[13] using a well-known process beginning with another chemical, β-ionone, as the raw material. This manufacturing process has been in use since the 1950s.

Given that, it is easy to see how the lines between "natural" and "synthetic" can become blurred and it demonstrates why they are meaningless. The β-carotene that is found in a banana is obviously a naturally-occurring component. But if this β-carotene was extracted from the banana, or any other food, and then used to color a colorless food, it can be called an artificial color. Butter is not yellow until β-carotene is added to it.

What is the verdict if the β-carotene that is used as a colorant came from a factory? Most people would probably say that would make it an artificial color, even though it is the same substance. They would probably be uncertain of the example where naturally-occurring β-carotene is added as an artificial color. So scientifically, how should the use of β-carotene in each of these three cases be characterized? Does it make the food artificially or naturally colored?

It's an irrelevant distinction, which is why these are marketing gimmicks rather than real issues. Chemically, it doesn't matter where β-carotene comes from, since neither the taste, smell, or any other properties are different. The origin of the chemical in this case, like in all cases, is meaningless, because the body cannot distinguish between synthetic β-carotene and β-carotene that is extracted from carrots.

The two are identical in every way, a concept that food scaremongers with no chemical expertise refuse to accept. The same concept holds true for banana flavor. Isoamyl acetate, aka "banana oil" is an acceptable artificial substitute for banana flavor, since it is the principal flavor found in bananas. A food that is flavored with isoamyl acetate will taste like banana, though it may lack a certain richness to some palates due to lacking the many other similar but subtly different flavors, much like the difference in flavors of wines.

Figure 1 illustrates how confusing this can be. Three separate banana bread recipes are shown, each with a subtle difference in the flavor ingredients. While the recipes A and C can easily be categorized as naturally flavored and artificially flavored, respectively, recipe B could be either, depending on vague and subjective criteria. But, more importantly, does it matter?

While the real answer is no, one could argue that it does matter, if you believe that all chemicals are carcinogens or health risks, because naturally-flavored bread could pose more of a health risk than an artificially-flavored counterpart, simply by virtue of it containing a greater variety of flavor chemicals.

But the hypothetical risk of natural banana bread constituting a real threat over an artificially-flavored kind is infinitesimally low. These flavors are only a few of the many thousands of chemicals that we ingest in varying quantities every day, be they natural or otherwise. Regardless of whether they are found in nature or synthesized in a lab, they have minimal toxicity, are ingested in minute quantities, or both. Additionally,

Figure 1
The blurred lines between artificial and natural flavoring

Are These Banana Breads
Naturally or Artificially Flavored?

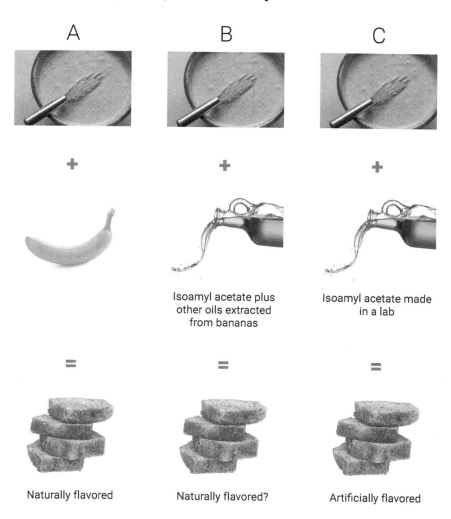

our bodies eliminate virtually all of the chemicals in our food rapidly. Why? The usual reason — we are biologically built that way.

Though it is surprising and counterintuitive to many consumers educated by natural foods marketing claims, when foods are artificially flavored they are still using the same chemicals that occur naturally, so there cannot be any difference in the health effects between the two. For example, if pure isoamyl acetate is used as a flavor substitute for bananas, the flavor of the food product in question may suffer, but the artificial flavor presents no additional health risk.

7
Exploitation of consumers by the "natural fallacy"

When bad science is promoted, it is reasonable to assume that there are economic benefits to be gained by those who are behind the scare-mongering. Not surprisingly, the food industry — both organic and, more recently, traditional — and aggressive environmental groups have this down to an art form.

Thanks to marketing campaigns that are anti-science at their core, the American public has been conditioned to equate artificial flavoring with harmful chemicals. Consumers are bombarded by terms such as "organic," "natural," and "synthetic" wherever they shop, without having anything close to a clear definition of what each term means. This loose, inconsistent use of terms may be on many labels, but the way they're used renders the information useless.

Today, the word "artificial" is a marketing death knell for products. Goods are being scrutinized more and more carefully as the fear of chemicals continues to grow. And that chemophobic framing against science certainly works.

According to Nielsen's January 2015 report "Healthy Eating Trends Around the World,[14]" more than 60 percent of Americans surveyed in 2014 considered the presence or absence of artificial flavors (and colors) to be an important consideration when selecting foods to purchase. This is largely due to manipulation by marketing organizations.

For example, Whole Foods uses a variety of tricks on its website. One, specifically, is a page titled "Unacceptable Ingredients for Foods," which is a long list of chemicals the company insists won't be included in the products it sells.

Whole Foods is lying when it claims that vanillin is not used in any of its products, stating "...[W]e won't sell a food product if it contains any of these." That's because the company omits that it simply won't sell a food product that has been flavored by *the addition of vanillin*. But Whole Foods is most certainly selling products that contain vanillin, and are flavored by it — anything that has a vanilla flavor. The company's trickery is based on the fact that it doesn't add vanillin. It is already there.

Whole Foods sells many vanilla products, and every single one of these contains *vanillin*, despite claims to the contrary. Without vanillin there can be no such thing as the flavor vanilla. The company is intentionally deceiving customers with claims that it "won't sell" a product containing vanillin when they simply mean they won't carry products to which vanillin has been *added*. Biologically, it doesn't make a bit of difference if vanilla products have been flavored by vanilla bean extract (roughly 80 percent vanillin) or vanillin that has been manufactured.

The company repeats this same marketing gimmick when it claims that a product does not have "added sugar," but rather has "evaporated cane juice." If you evaporate sugar cane juice, you know what you are left with and so do they: sugar. Evaporated cane juice *is* sugar, Whole Foods just gives it a different name to make it sound healthier than added sugar.

Since Whole Foods has a page titled "Unacceptable Ingredients for Foods" and states "we wont sell a food product if it contains any of these" but they really do, it's reasonable to ask what the purpose of the list really is. Is Whole Foods implying that there's a health risk associated with these chemicals unless they are present naturally or change the name? It must remain a marketing mystery, but the company is not pulling vanilla products off of their shelves.

8
Summary

Although this book is about science, it could just as well be found in the psychology or business sections of a bookstore. Psychology plus business equals marketing. But in the case of food, marketing is based on unfounded fear and misconception of chemicals — an often-used but nonetheless exceedingly successful tactic against science.

How could it be otherwise? Minute or even nonexistent threats from chemicals have become so central to our modern core belief system that no amount of education seems to be able to shake it. Even in a supposedly educated country, it is quite easy to find people who cannot, or will not, comprehend that all matter is made of chemicals. Further, they continue to believe that chemicals somehow have different properties depending on whether they come from a house plant or a manufacturing plant.

The food industry is as guilty as any of perpetuating this myth. The pipe-dream of the absence of chemicals in our lives now just "sounds right" to consumers. $100 billion in consumer spending shows many readily buy into the chemical-free, organic mentality that has overtaken purchasing decisions.

On some level, who can blame them? People are frightened by what they don't understand, and so few people have even a marginal knowledge of chemistry that the word "chemical" itself has taken on a pejorative meaning, despite the fact that life depends on the very chemicals that many people fear.

So it's no surprise that artificial flavors get a bad rap. Artificial means "chemical" and "chemical" means unhealthy — something that those who market many products, especially foods, know all too well. While there's no direct harm when a consumer chooses a naturally-flavored food over one that is artificially flavored, there is indirect harm, both in terms of paying higher prices for an item with little or no added benefit, and the further erosion of science-based thinking in the country.

In the 21st century, acceptance of science is a real concern, not an artificial one.

References

1. "7 Worst Ingredients in Food" October 18, 2012, http://bit.ly/2gGdVwV; Accessed 11/22/16

2. Gaffney, S.H., et. al "Naturally occurring diacetyl and 2,3-pentanedione concentrations associated with roasting and grinding unflavored coffee beans in a commercial setting." *Toxicology Reports*, 2, 1171-1181 (2015).

3. Buttered Popcorn Flavoring Linked To Alzheimer's, October 18, 2012 http://bit.ly/2gGdVwV Accessed 11/22/16

4. Ranadive, A. "Vanillin and related flavor compounds in vanilla extracts made from beans of various global origins" *J. Agric. Food Chem.*, 40 (10), 1922–1924, (1992).

5. 4-Hydroxybenzaldehyde MSDS—Sigma-Aldrich Data Catalogue. Version 3.7. 11/13/16. http://bit.ly/2gy8sHq; Accessed 11/22/16

6. 4-Hydroxybenzoic acid MSDS—Sigma-Aldrich Data Catalogue. Version 5.4. 11/13/16. http://bit.ly/2gcrhvL; Accessed 11/22/16

7. Vanillic acid MSDS —Sigma-Aldrich Data Catalogue. Version 3.8. 11/14/16. http://bit.ly/2fBFHW4; Accessed 11/22/16

8. Dharmadhikari, M. "Composition of Grapes" http://bit.ly/2gdYVF1 Accessed 11/22/16

9. Washburn, S. "How Many Different Types of Wine Grapes Are There?" *Wine Guide.* http://bit.ly/2flFQMK. Accessed 11/22/16

10. Williams, P.J., Strauss, C.R., Wilson, B. "Hydroxylated linalool derivatives as precursors of volatile monoterpenes of muscat grapes." *Journal of Agricultural Food Chemistry*, 28(4), 766-771 (1980).

11. "Beta-Carotene. Nutri-Facts: Understanding Vitamins & More" http://bit.ly/2fnvr6c. Accessed 11/22/16

12. Ribeiro, B.D., Barreto, D.W. & Coelho, M.A.Z. (2011) Technological Aspects of β-Carotene Production. *Food Bioprocess Technol* 4(5): 693-701.

13. "Nielsen Global Health and Wellness Report" http://bit.ly/2flXDmX; Accessed 11/22/16

14. Whole Foods "Unacceptable ingredients for foods" http://www.wholefoodsmarket.com/about-our-products/quality-standards/food-ingredient; Accessed 11/22/16

The opinions expressed in ACSH publications do not necessarily represent the views of all members of the ACSH Board of Trustees, Founders Circle and Board of Scientific and Policy Advisors, who all serve without compensation.

Printed in Great
Britain
by Amazon

31055870R00027